Opens February 20

PANAMA · PACIFIC
INTERNATIONAL
· EXPOSITION ·
SAN FRANCISCO · 1915

Thirteenth Labor of Hercules

Introductory

N FEBRUARY 20, 1915—on time and minutely ready—the Panama-Pacific International Exposition will be opened at San Francisco.

It will be the third exposition of its class held in the United States and the twelfth of its class held anywhere in the world. It is the official, national and international celebration of a contemporaneous event—the opening of the Panama Canal.

The propriety of celebrating that event by a great Universal Exposition was recognized several years ago. It was recognized at the same time that as the event was of transcendent importance its celebration, to be adequate, must transcend all precedent. If the greatest physical achievement in history was to be celebrated by an Exposition, then that Exposition should be the greatest in history.

Only those who have seen and studied the great universal expositions of the past can realize the task involved in building one that should excel them all. Great as this task appeared, the honor of assuming it was sought by many cities. After much consideration Congress, in 1910, entrusted the responsibility to San Francisco and the Panama-Pacific International Exposition represents the fulfillment of that national trust.

Such is the genesis, briefly stated, of the great Exposition to which California, as the hostess-state, invites the world in 1915.

It seems particularly fitting that this invitation should come from California. It seems fitting that the Exposition, which marks the beginning of a new era in commerce, should be held on the shores of the Pacific. California marks the limit of the geographical progress of civilization. For unnumbered centuries the course of empire has been steadily to the west. On the shores of the Pacific it finds itself still facing west, yet looking to the east; or, in Whitman's beautiful phrasing:

"Facing west, from California's shores,
Inquiring, tireless, seeking what is yet unfound,
I, a child, very old, over waves, towards the house of
   maternity, the land of migrations, look afar,
Look off the shores of my Western Sea—the circle
   almost circled."

This Exposition therefore marks the beginning of a new era in civilization. The circle is now fully circled; the West has met the East.

There can be no cessation of the progress of mankind, but as that progress can no longer be, in the geographical sense, onward, it must be, in the ethical sense, upward.

Geographically, the Exposition is fittingly placed on the shores of the Pacific, because of the new and immense importance which the nations of the Pacific area, under the stimulus of the Panama Canal, will now assume in the eyes of commerce.

And such is the practical advantage offered by California as a state for the reception of Exposition visitors. The climate of San Francisco permits the Exposition to remain open for ten months, offering a temperature unchangingly comfortable during the whole period.

In this garden of the earth, at the great Panama-Pacific International Exposition, in 1915, man will meet his fellows from the four quarters of the globe. There will be free expression of thought, a comparison of methods, and an interchange of ideas such as the world has never known. And this is the greatest purpose of all great expositions. They infallibly broaden the mental horizon of the individual visitor and thereby lead to greater social sympathies, to the harmonizing of geographical viewpoints, to better national understandings.

Entirely aside from the practical instruction to be had from the commercial, scientific and educational exhibits, no one can visit San Francisco and the great Universal Exposition of 1915 and fail to receive in addition a mental, social and spiritual stimulus.

The Event

Cutting the Panama Canal

SOME four centuries ago, Balboa, the intrepid, the persevering, led his little band of adventurers across the Isthmus of Darien, as it was then called, and leaving their protection, gave rein to his impatience by going on ahead and climbing alone, slowly and painfully, the continental divide, from which vantage point he discovered the world's largest ocean.

We are told that, later, gathering his followers, he walked out into the surf and with his sword in his right hand and the banner of Castile in his left he gave the vast expanse of water its present name and claimed all the land washed by its waves as the lawful property of the proud country to which he owed allegiance.

The narrowness of the Isthmus naturally suggested the cutting of a waterway through it. It interposed between Atlantic and Pacific a barrier in places less than fifty miles wide. To sail from Colon to Panama—forty-five miles as the bird flies—required a voyage around Cape Horn—some ten thousand miles. Yet it was nearly four centuries before any actual effort was made to construct such canal.

In 1876 an organization was perfected in France for making surveys and collecting data on which to base the construction of a canal across

City of Colon, Atlantic Side of Canal

the Isthmus of Panama, and in 1878 a concession for prosecuting the work was secured from the Colombian Government.

In May, 1879, an international congress was convened, under the auspices of Ferdinand de Lesseps, to consider the question of the best location and plan of the canal.

The Panama Canal Company was organized, with Ferdinand de Lesseps as its President. The stock of this company was successfully floated in December, 1880. The two years following were devoted largely to surveys, examinations and preliminary work.

In 1889 the company went into bankruptcy, and operations were suspended until the new Panama Canal Company was organized in 1894.

The United States, not unmindful of the advantages of an Isthmian Canal, had from time to time made surveys of the various routes. With a view to Government ownership and control, Congress directed an investigation, with the result that the Commission reported on November 16th, 1901, in favor of Panama and recommended the lock type of canal, appraising the value of the rights, franchises, concessions, lands, unfinished work, plans and other property, including the railroad of the new Panama Canal Company, at $40,000,000. An act of Congress approved June 28th, 1902, authorized the President of the United States to acquire this property at this figure, and also to secure from the Republic of Colombia perpetual control of a strip of land not less than six miles wide across the Isthmus and the right to excavate, construct and operate and protect thereon a canal of such depth and capacity as will afford convenient passage of the largest ships now in use or as may be reasonably anticipated.

Later on a treaty was made with the Republic of Panama whereby the United States was granted control of a ten-mile strip constituting the

The Great Fourteen-Gate Spillway at Gatun

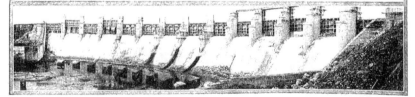

Pedro Miguel Locks   Water in Culebra Cut

Canal Zone. This was ratified by the Republic of Panama on December 2nd, 1903, and by the United States on February 23rd, 1904.

On May 4th, 1904, work was begun under United States control.

The Isthmus of Panama runs east and west and the canal traverses it from Colon on the north to Panama on the south in a general direction from northwest to southeast, the Pacific terminus being twenty-two miles east of the Atlantic entrance.

The principal features of the canal are a sea-level entrance channel from the east through Limon Bay to Gatun, about seven miles long, 500 feet bottom width and 41 feet deep at mean tide. At Gatun the 85-foot lake level is obtained by a dam across the valley. The lake is confined on the Pacific side by a dam between the hills at Pedro Miguel, 32 miles away. The lake thus formed will have an area of 164 square miles and a channel depth of not less than 45 feet at normal stage.

At Gatun ships will pass from the sea to the lake level, and vice versa, by three locks in flight. On the Pacific side there will be one lowering of 30 feet at Pedro Miguel to a small lake 55 feet above sea level, held by dam at Miraflores, where two lowerings overcome the difference of level to the sea. The channel between the locks on the Pacific side will be 500 feet wide at the bottom and 45 feet deep, and below the Miraflores locks the sea-level section, about eight miles in length, will be 500 feet wide at the bottom and 45 feet deep at mean tide. Through the lake the bottom widths are not less than 1000 feet for about 16 miles, 800 feet for about 4 miles, 500 feet for about 3 miles, and through the continental divide from Bas Obispo to Pedro Miguel, a distance of about 9 miles, the bottom width is 300 feet.

The total length of the canal from deep water in the Caribbean, 41-foot depth at mean tide to deep water in the Pacific, 45-foot depth at mean tide is practically 50 miles, 15 miles of which are at sea level.

Culebra Cut, Showing the Cucaracha Slide

Emergency Dams each side of Gatun Locks

The greatest difficulty encountered in the excavation of the canal has been due to slides and breaks which caused large masses of material to slide or move into the excavated area, closing off the drainage, upsetting steam shovels and tearing up the tracks. The greatest slide was at Cucaracha, and gave trouble when the French first began cutting in 1884. Though at first confined to a length of 800 feet, the slide extended to include the entire basin south of Gold Hill, or a length of about 3000 feet. Some idea of the magnitude of these slides can be obtained from the fact that during the fiscal year 1910 of 14,921,750 cubic yards that were removed 2,649,000 cubic yards, or 18 per cent, were from slides or breaks that had previously existed or that had developed during the year.

The one greatest undertaking of the whole excavation is the Culebra cut. Work has been in progress on this since 1880, and during the French control over 20,000,000 cubic yards were removed. On May 4th, 1904, when the United States took charge, it was estimated that there was left to excavate 150,000,000 cubic yards. Some idea of the size of this big cut may be formed from the fact that this division has within its jurisdiction over 200 miles of 5-foot-gauge track laid, about 55 miles of which is within the side slopes of the Culebra cut alone.

The great dam at Gatun is a veritable hill—7500 feet over all, 2100 feet wide at the base, 398 feet through at the water surface, and 100 feet wide at the top, which is 115 feet above sea level. The dimensions of the dam are such as to assure that ample provision is made against every force which may affect its safety, and while it is made of dirt, a thing before unheard of, it is of such vast proportions that it is as strong and firm as the everlasting hills themselves.

Fluctuations in the lake due to floods are to be controlled by an immense spillway dam built of concrete. The front of the dam is the arc of a circle 740 feet long with 14 openings which, when the gates are raised to the full height, will permit a discharge of 140,000 cubic

Cucaracha Slide, which carried away the railroad

The Entrance of Gatun Locks from Gatun Lake

feet per second. The water thus discharged will pass through a diversion channel in the old bed of the Chagres River, generating, by an enormous electric plant, the power necessary for operating the locks.

The locks of the canal are in pairs, so that if any lock is out of service navigation will not be interrupted; also, when all the locks are in use the passage of shipping will be expedited by using one set of locks for the ascent and the other for descent. These locks are 110 feet wide and have usable lengths of 1000 feet. The system of filling adopted consists of a culvert in each side wall feeding laterals perpendicular to the access of the lock, from which are openings upward into the lock chamber. The entire lock can be filled or emptied in fifteen minutes and forty-two seconds when one culvert is used and seven minutes and fifty-one seconds using both culverts. It is estimated that it will require about ten hours for a large ship to make the entire trip through the canal.

Many extraordinary feats of engineering were accomplished to overcome the difficulties presented. Special contrivances, wonderful in their operation, were invented to meet exigencies and emergencies.

The first and greatest problem attempted by the United States was how to make the Canal Zone healthful. This strip of land from ocean to ocean abounded in disease-breeding swamps and filthy habitations unfit for human beings. The death rate was appalling and the labor conditions terrible.

During the first two and a half years all energies were devoted to ridding the Isthmus of disease by sanitation, to recruiting and organizing a working force and providing for it suitable houses, hotels, messes, kitchens and an adequate food supply. This work included clearing lands, draining and filling pools and swamps for the extermination of the mosquito, the establishment of hospitals for the care of the sick and injured and the building of suitable quarantine quarters.

Miraflores Locks, Lower Level

Blasting Gamboa Dyke   The Meeting of the Waters

Municipal improvements were undertaken in Panama and Colon and the various settlements in the Canal Zone, such as the construction of reservoirs, pavements and a system of modern roads. Over 2000 buildings were constructed beside the remodeling of 1500 buildings turned over by the French company.

It was only after all this preliminary sanitation was accomplished that the real work of digging the canal could go forward with any hope of success. These hygienic conditions had the result of making the Canal Zone one of the most healthful spots in the world and work on the canal became so popular that it was no longer necessary to enlist recruits from the West Indies, the good pay, fair treatment and excellent living conditions bringing thousands of laborers from Spain and Italy. The greatest number employed at any one time was 45,000, of which 5000 were Americans.

The completion of this herculean task marks an epoch in the history of the world. A gigantic battle against floods and torrents, pestilence and swamps, tropical rivers, jungles and rock-ribbed mountains has been fought—and WON!

Well worthy a place in the halls of immortal fame are the names of the thousands of sturdy sons who with ingenuity, pluck and perseverance never before equaled have succeeded in making a pathway for the nations of the world from ocean to ocean.

This great and daring undertaking, which had for its object the opening up of new trade routes and lines of commerce, annihilating distance and wiping out the width of two continents between New York and Yokohama and making the Atlantic seaboard and the Pacific Coast close neighbors, is the climax of man's achievement and the greatest gift to civilization. It will help in the consummation of man's loftiest dreams of world friendship and world peace.

Cathedral Square, City of Panama, Pacific End of Canal

The Celebration

The Panama-Pacific International Exposition

FITTINGLY commensurate with the heroic achievement will be the celebration in its honor.

The completion of the Panama Canal being an accomplishment affecting the interests of every civilized nation, the celebration thereof naturally suggested was a great Universal Exposition in which all nations should participate under the auspices of the United States. As already mentioned in the introduction to this booklet, Congress designated San Francisco as the place for such an Exposition and entrusted to that city and to the State of California the responsibility of providing for the reception of the nations of the world and for the housing of the exhibits which should best demonstrate their achievements.

This responsibility was gladly accepted by California and the work of carrying out the duty to the nation was begun immediately.

In the space of a few days a fund approximating $20,000,000 was raised by the citizens of San Francisco, the municipality and the State Legislature. The fifty-eight counties of California are raising several million dollars for their individual displays, which will be on an elaborate scale. This will be added to by the various states throughout the Union and materially augmented by liberal amounts from foreign countries the world over. It has been conservatively estimated that the grand total will represent an expenditure exceeding $50,000,000.

Over the Domes, through the Golden Gate

Palace of Education, May 1, 1914

May Day Exercises, Exposition Site

In magnificence and splendor, number of palaces, beauty of grounds, number and quality of exhibits, diversity of subjects, completeness of detail and hugeness of the whole this will be an exposition adequate to the event it celebrates. It will have great and lasting effect upon the trade, relationships and commercial activity of all countries.

The Panama-Pacific International Exposition sets a new standard for world expositions:

It is *universal* in plan in that it includes the peoples and products from every section of the globe, giving invitation to all alike and making every effort to have all their resources and achievements represented by exhibits. In brief, it will be a cross-section of human accomplishment.

It is *contemporaneous* in character. Nothing will be shown or demonstrated in competition for award that does not have its place in our every-day life. While examples of the civilization of all ages will be shown for the purpose of comparison and education, the results of the achievements of the present decade will have the place of honor.

It is *selective*. This point alone guarantees a degree of quality that could not be obtained in any other way. In this Exposition representative types will have the first consideration. All exhibits will go through a vigorous process of investigation, examination and elimination before they are finally approved. This is the most valuable feature of the Exposition from the viewpoint of both the exhibitor and the visitor.

The Exposition site combines to an extraordinary degree the qualities of beauty, adaptability and convenience. It is a natural amphitheatre, fronting on the wonderful island-dotted Bay of San Francisco, just inside the famous Golden Gate. Towering, wooded heights flank it at each end, while at its back the hills roll up sharply. All this is in the very heart

Dedicating the Idaho Building

Crowds in Palace of Machinery

of the best residential district of San Francisco and within fifteen minutes street car ride from the City Hall.

The view facing north is across the sparkling waters of the Bay—a land-locked harbor which resembles a great mountain lake more than an arm of the sea. It is rimmed around by mountains, with haughty Tamalpais towering over all. Under the turquoise California sky, drenched with sunshine and color, it is a scene at once exquisitely beautiful and solemnly majestic.

With this wonderful scene as a background, the architects, artists and landscape gardeners of the Exposition have planned and erected a city straight out of a beautiful dream. It can not be described in words nor adequately shown in picture, although an attempt to give the reader an idea of its general color effect appears elsewhere in this booklet.

The proximity of the site to a world waterway is a wonderful advantage. Through the portals of the Golden Gate the nations of the earth can bring their richest offerings to the very gates of the Exposition, avoiding a long continental haul and consequent damage from reshipping.

The site adapts itself to the carrying out of wonderful aquatic displays. Carnivals, maneuvers by the fleets of all nations, international yacht racing, motor boat racing, exhibitions by submarines and hydroplanes, all can be indulged in in the immediate foreground of the Exposition palaces. The entire navies of the world can here assemble and land their crews right on the edge of the beautiful "Marina."

The main exhibit palaces, eleven in number, contain, under a comprehensive and representative classification, examples of the resources and achievements along all lines of human endeavor, which are divided into departments as follows: "A"—Fine Arts; "B"—Education; "C"—Social Economy; "D"—Liberal Arts; "E"—Manufactures and Varied Indus-

The Million Dollar Exposition Auditorium in the Civic Center

The Panama-Pacific International Exposition Site, April 1, 1913

tries; "F"—Machinery; "G"—Transportation; "H"—Agriculture; "I"—Agric. (food products); "K"—Horticulture; "L"—Mines and Metallurgy.

These eleven great palaces, together with Festival Hall, form the central setting of a beautiful picture, flanked on the city side by the amusement section or concessions district and on the other end by the buildings of the various states and the pavilions of the foreign nations. These latter join the aviation field, race track and live stock exhibit, terminating in the grounds of a great military reservation, the Presidio, where the competitive drills and army maneuvers will take place.

In formation the eight main exhibit palaces—Education, Liberal Arts, Manufactures, Varied Industries, Agriculture, Food Products, Transportation and Mines and Metallurgy—represent a quadrangle, being bisected by an avenue east and west and intersected by avenues north and south, the intersections marking the three great courts. The facades of the palaces are the walls of these courts and partake of the particular style of architecture dominating the court on which they front. These eight palaces are flanked on the east by the great Palace of Machinery and on the west by the Palace of Fine Arts.

Passing through the main gate on the city side the visitor enters the great South Garden, 3000 feet in length, on the right extremity of which can be seen the beautiful Festival Hall. To the extreme left is the Palace of Horticulture. Immediately in front is the Main Tower or "Tower of Jewels." This great garden, itself a marvel of landscape engineering skill, is but one side of a magic carpet on which these beautiful palaces are set, the 300-foot wide "Marina" and its grand esplanade, with its floricultural splendors, forming the other side, the pattern threading its winding way through the various courts and recesses over the entire grounds, forming a correlated whole which, for wondrous beauty, has never been equaled.

Panorama of the Site Showing the

The Panama-Pacific International Exposition Site, April 1, 1914

Passing from this great garden under the arch of the main tower the visitor enters the "Court of the Universe," the largest of the five courts of the Exposition. This is the meeting place of the Eastern and Western hemispheres, and the decorative scheme on each side is typical of this theme. On the extreme right and left are two great Triumphal Arches, the one on the right, which leads to the Court of Abundance, being surmounted by a magnificent statuary group, "The Nations of the East," the figures symbolizing life in the Orient, while the arch on the left, leading to the Court of the Four Seasons, has a group of the same proportions, "The Nations of the West," symbolical of life in the Occident. Straight ahead is the colossal column of Progress, surmounted by the "Adventurous Bowman" shooting the arrow toward the West.

To the right, under the "Arch of the Rising Sun," is the avenue leading to the "Court of Abundance," which terminates at its southern extremity into the "Court of Flowers," one of the minor courts; while to the left, under the "Arch of the Setting Sun," is the avenue leading to the beautiful "Court of the Four Seasons," which, at its southern extremity, enters the other minor court, the "Court of Palms."

Continuing straight ahead one comes to the edge of the spacious Yacht Harbor and the center of the Grand Esplanade or "Marina." Long after the Exposition is over—when it is only a fond and loving memory —this esplanade will remain to grace and enhance the natural beauties of San Francisco Bay.

While color will be the dominant note—color grouped in large masses of reds, blues, greens and golds—yet over all will prevail harmony, the palaces themselves being of a soft, neutral tint—a smoked ivory—that is at once pleasing and restful to the eyes.

One of the most attractive and beautiful features of this Exposition will be the electrical illumination. By an entirely new system of flood

Main Exhibit Palaces, May 1, 1914

Looking Across the "Concessions" District to the Marin Hills

lighting a soft, restful, yet perfect light will pervade the courts at night, revealing in wonderful clearness the facades and walls of the palaces and the natural colors of the shrubbery and flowers. By peculiar and novel lighting devices the statuary and mural paintings will be made to appear with even heightened effect. Concealed batteries will project powerful yet softened rays of light that will cause tens of thousands of specially prepared glass "jewels," hung tremulously upon the towers, to flash and scintillate like great diamonds, emeralds and rubies. At a point on the bay shore will be erected apparatus that will weave in the night sky auroras of ever-changing color. Altogether the spectacle will be interesting and wonderful and never to be forgotten.

Exposition Palaces and Courts

Palace of Fine Arts. Designed by R. B. Maybeck. Its length from north to south describes an arc eleven hundred feet. This palace will face upon a great lagoon of placid water which will reflect its beautiful architecture. It will be a fire-proof structure. In the center of the arc will be erected a great dome with steps leading down to the lagoon in a beautiful setting of shrubbery, composed of Monterey Cypress and other evergreen trees, making perhaps the prettiest setting of the whole Exposition site. The painting and sculpture of every nation of artistic prominence will be artistically shown in this palace. The exhibits in the United States section will consist not only of the work of contemporary artists, but of historic American paintings from the time of West, Copley and Stuart to the present and a loan collection of canvases by foreign artists owned in the United States. The installation of the canvases and small bronzes will be intimate. The color scheme of the galleries varying to serve as a sympathetic background for their contents.

Palace of Horticulture. Designed by Messrs. Bakewell & Brown of San Francisco. This palace is constructed almost entirely of glass and covers over five acres. It is surmounted by a dome 160 feet in height.

Battery of Steam Rollers, Making Perfect Roads, May 1, 1914

CROWNING GROUPS, COURT OF THE UNIVERSE

Nations of the West
CALDER - LENTELLI - ROTH - SCULPTORS

Nations of the East
CALDER - LENTELLI - ROTH - SCULPTORS

Two Views of the Court of the Four Seasons

It is 672 feet long and its greatest width is 320 feet. An imposing nave 80 feet in height runs the length of the building and paralleling the central nave are (one on either side) two side aisles each 50 feet in height. All phases of practical horticulture will be embraced in this exhibit. Among other things a fully equipped fruit-canning establishment will be in operation, showing the sanitary way in which fruit is prepared and canned; a seed-packing establishment, orange-packing house, olive oil presses in operation; tools used in the culture of fruits, trees and flowers. The frostless climate of California, which enables plant life to attain the highest perfection, will give the floricultural exhibit a distinction and beauty it has not been possible to attain at other expositions where the seasons have been short and the winters severe. There is to be a rose contest in which the Exposition offers as trophy a thousand-dollar cup to the originator of the finest new seedling rose which has never before been exhibited.

The Great Palace of Machinery. Designed by Messrs. Ward & Blohme of San Francisco. This palace is the largest building erected on the Exposition site. It is 968 feet by 368 feet. One mile and a half of cornices was used in ornamenting the building. Four carloads of nails and 1500 tons of bolts and washers were used in its construction. In this palace will be assembled exhibits of machinery used in the generation, transmission and application of power. Several groups will comprise examples of steam generators and motors utilizing steam, internal combustion motors, hydraulic motors, miscellaneous motors, general machinery apparatus and accessories, and tools for shaping wood and metals. Ten special electrical groups will cover the generation, distribution and control of electrical energy in its application to mechanical and motor power, lighting and heating.

It is not possible to name any one person as the designer of the following eight palaces comprising the central group. They are composite in design, each facade partaking of the particular style of architecture dominating the court on which it fronts.

Palace of Education and Social Economy. The exhibits in this palace will show development along these lines since 1905, and by specializing on

The Palace of Machinery Compared in Size to a Railroad Train

CELEBRATING
THE
OPENING
OF THE

Two Views of the Avenue of Progress. May 15, 1914

prominent movements and reforms will seek to forecast the education of tomorrow. There will be a comparative exhibit of the educational systems of all nations participating and a comprehensive demonstration of educational work in the United States in all its phases from kindergarten to university. The Department of Social Economy will bring together a comprehensive collection of exhibits illustrative of the conditions and necessities of man considered as a member of organized society and government, together with displays showing the agencies or means employed for his well being. As far as possible, operating examples will be given. Child welfare, and the work of organizations such as boy scouts, campfire girls, etc., charities, corrections, criminology, urban problems, park systems, public buildings, street improvements, method of disposing of sewage, etc., will receive exhaustive treatment by exhibits. Such matters as finance in its relation to the public welfare and in connection with such agencies as banks and provident associations, modern credit associations, etc., will be illustrated. All matters pertaining to commerce in the way of distribution of goods, business standards and systems; all labor problems involving working conditions and standards, welfare and efficiency, and including domestic science and woman's vocations, will be exhaustively studied and compared by exhibits. The latest discoveries in hygiene, methods of missionary work, international and universal peace institutions, diplomatic and consular systems—all these will receive a broad and sympathetic treatment by exhibitional studies.

Palace of Liberal Arts. Liberal Arts rank high in the classification of exhibits because they embrace the applied sciences which indicate the result of man's education and culture, illustrate his tastes and demonstrate his inventive genius and scientific attainment and express his artistic nature. This splendid palace is directly opposite the main entrance to the Exposition grounds from the city side and is approximately 585 feet long, 470 feet wide and 65 feet high and covers nearly six acres.

Palaces of Manufactures and Varied Industries. The department of a universal Exposition which has the combined interest of all nations is the exhibition of finished products of manufacture and manual skill, the objects of utility, luxury and taste in which each country excels and

South Facade Education, May 15, 1914   Mines and Metallurgy, May 15, 1914

Palace of Machinery   May 15, 1914   Palace of Horticulture

which constitute the most valuable and profitable part of foreign trade. The various nations will show with honest pride and satisfaction their artistic products. The art industries of the world will be brilliantly displayed in the Palaces of Manufactures and Varied Industries.

Palace of Transportation. The exhibits in this palace will be made, as far as possible, contemporaneous, not historical. There will be displayed the very latest achievements of human ingenuity covering the entire field of transportation. On account of the great development of the motor boat industry and aerial navigation these two groups will be thoroughly represented, both in indoor and outdoor exhibits. In this palace will be shown the exhibits of all the great steamship companies, the water transportation of all countries, their navigation and commerce, characteristic boats and ships of all nations. Sail and steam yachts will be generally shown by models. Electric exhibits will show the latest application of electricity to the agency of transportation. A locomotive exhibit will illustrate the latest types. Car exhibits will show the modern development of street car equipment, and there will be a complete showing of railway supplies, including all the new inventions and appliances for the protection of life and property in this connection.

Palace of Agriculture. The section of this Exposition devoted to the interests of agriculture will embrace an area of more than forty acres. The Palace of Agriculture, proper, will cover seven and one-half acres. The exhibits will deal with every possible phase of the agricultural industry. A very important group will be devoted to farm implements and machinery. No less than seven distinct classes will be required. In this department also will be shown all that pertains to forestry and forest products.

Palace of Agriculture (food products). Under this same department, although in a separate palace, the multiform exhibits governing the food products of the entire world will be grouped. Vegetable and animal food products and the equipment and methods employed in the preparation of foods and beverages will be extensively shown.

Palace of Mines and Metallurgy. The exhibits in this palace deal with the natural mineral resources of the world, their exploration and

Palace of Varied Industries, May 15, 1914

Philosopher

Star Figure

The Priest

Sunshine

Rain

Autumn

SELECTIONS FROM
THE MANY BEAUTIFUL
SCULPTURAL SUBJECTS

Summer

Spring

Ferry Slip, Exposition Grounds   One of the Fire Stations

exploitation, their conversion into metal, their manufacture into structural forms and into raw material for the various industries. They take in the ordinary metallics such as gold, silver, copper, lead, zinc, iron, aluminum, etc.; the rare metallics such as tungsten, vanadium, uranium, radium, platinum, etc.; the non-metallics, such as clay, cement and their products; coal, oil and gas; the salines, fertilizers, etc. The object of the Mines and Metallurgy exhibit is two-fold: first, to draw attention to the natural mineral resources of each country, state or community so that the public may learn of the mode of occurrence of the metals of commerce and their distribution, of the stage of development of the various districts, of present sources of supply and consumption and of possible future sources of supply and of extended markets; second, to educate the public in a general way regarding the details of the industry, its problems and its needs.

Live Stock Exhibit. In keeping with the general plan of the Exposition the Department of Live Stock will be presented in a better manner than has heretofore characterized such exhibitions. Competitions for the $175,000 in prize money appropriated by the Exposition, and for the supplemental premiums offered by the breeders' associations will take place in the months of October and November. In addition to this there will be a continuous live stock display in 1915 from February 20th to December 4th. In housing, classification and arrangements of the exhibits, the Department of Live Stock at San Francisco will demonstrate the advancement that has been made since the last world exposition. Special events will include universal polo, international cavalry contests, two harness horse racing meets, and the carrying on of a series of demonstrations which will teach everything that is new in this important industry.

The Exposition Auditorium. Designed by Messrs. John Galen Howard, Frederick H. Meyer and John Reid, Jr. The Exposition Auditorium will be a four-story construction of steel and stone and will grace the Civic Center of San Francisco. It will be a lasting and beautiful monument to the Panama-Pacific International Exposition. The Exposition management is paying over one million dollars for its erection and the City and County of San Francisco is paying nearly a million dollars for the site.

Palace of Horticulture   May 25, 1914   Vestibule, Palace of Machinery

Removing and Transplanting Large Palms on Exposition Grounds

The main auditorium of this building will accommodate twelve thousand persons.

Festival Hall. Designed by Robert Farquhar of Los Angeles. This will be the scene of many of the great festivals and choral competitions entered into by the various singing organizations of the world. Festival Hall is built in the French theatre style of architecture with one large dome and various minor domes and minarets, profusely decorated with statuary. The main hall will contain seats for about three thousand persons, and here will be placed a huge pipe organ which is seventh in size in the world.

The California Building. Designed by Thos. H. F. Burditte. This building will be in the old Mission style and will cover approximately 350 feet by 675 feet. In form it will consist of a towered main building, two stories in height and surrounded by an immense court. Its construction and furnishings will represent an outlay of considerably over half a million dollars. This will be the "Host Building" of the Exposition. It will contain the displays of the fifty-eight counties of California. This building with its walled-in court and park will cover about seven acres. The Woman's Board, an auxiliary of the Exposition, has assumed the responsibility of furnishing and maintenance, and will have entire charge of its social administration.

Main Tower or "Tower of Jewels." Designed by Messrs. Carrere & Hastings of New York. This tower will rise to a height of 433 feet and, from an architectural standpoint, will be the dominating feature of the Exposition. This will be the center of a brilliant night illumination, the outline of the tower being defined by over one hundred thousand hand-cut glass "jewels" or prisms, hung tremulously, the least atmospheric disturbance causing them to flash and change and scintillate in a thousand different tints and colors.

The Court of the Universe. Designed by Messrs. McKim, Mead and White of New York. This is the great central court or court of honor of the Exposition, and in design and decoration it is made to represent the meeting place of the hemispheres. It will be 700 feet long and 900 feet wide, and will contain a sunken garden in the center. At the northern

Dedicating California Building   Autos Waiting for the Crowds

Main Portal, Varied Industries   Minor Portal, Varied Industries

end between the Palaces of Agriculture and Transportation will be a great pool of water embellished with statuary and fountains.

The Court of Abundance. Designed by Louis C. Mullgardt. This is the east central court of the Exposition and in design will show the Oriental phase of the Spanish-Moorish type. This court will be dedicated to music, dancing, acting and pageantry.

The Court of the Four Seasons. Designed by Henry Bacon of New York. This is the west central court and one of the most beautiful sections of the Exposition. It is said that Hadrian's Villa, one of the historic Roman palaces, is the inspiration for this court. It will be surrounded by a beautiful colonnade, in each of the four corners of which will be niches containing statuary representing the four seasons.

The Court of Palms. Designed by George W. Kelham of San Francisco. This is one of the two minor courts of the Exposition. Its entrance is from the great South Garden between two towers, each rising to a height of 200 feet and favoring in architecture the period of the Italian Renaissance. This court will contain a showing of rare and beautiful palms.

The Court of Flowers. Designed by George W. Kelham of San Francisco. This is the second of the minor courts, also having its entrance from the great South Garden between two Italian towers almost the exact duplicate of those at the entrance to the Court of Palms. While being the smallest of the Exposition courts it will be nevertheless as beautiful as the others and, as the name denotes, will be a paradise of vari-colored flowers.

The Amusement or Concessions District. "The Zone," the main amusement street, is 3000 feet in length and runs through the center of this district. Sixty-five acres is devoted to the amusement features of the Exposition. This section will be open from nine-thirty a. m. until eleven o'clock p. m. When this division of the Exposition is opened it will represent an outlay of over ten million dollars. Rigid selection has governed the granting of all of the concessions. Every one admitted has satisfied a high standard of propriety, good taste, and educational value, as well as effective fun-making and entertainment.

California Building, Which Will Contain the Displays of Its Fifty-eight Counties

Inside the Court, California Building

On the Steps of the Service Building

Statuary. Over 250 distinct groups and hundreds of individual pieces of statuary will be shown at the Panama-Pacific International Exposition. These comprise among others the following groups and subjects: "Nations of the East," "Nations of the West," "The Colossal Column of Progress," "Spring," "Summer," "Autumn," "Winter," "Fountain of Energy," "Fountain of Youth," "The Fountains of the Rising and Setting Sun," "Fire," "Water," "Earth" and "Air," "Order and Chaos" and "Eternity and Change," "Modern Civilization," "Armored Horseman," "Philosopher," "Adventurer," "Priest," "Soldier," "Fountain of Eldorado," "Nature," "Ceres," "Beauty and the Beast," "End of the Trail," "The Pioneer," "Cortez," "Pizarro," "The Miner," "The Pirate," "Primitive Man," "Primitive Woman," "Steam," "Electric Power." Besides these figures and groups are many beautiful friezes, spandrels, capitals, niches and columns decorated with allegorical subjects. This array will combine to make statuary at the Panama-Pacific International Exposition one of the leading and most interesting features.

Progress. The progress of the Panama-Pacific International Exposition is unparalleled in the history of expositions and is a forecast of the magnificence and comprehensiveness of the Exposition upon its opening day, February 20 of 1915. The progress of the construction work is visible to every visitor to the grounds. The eight main exhibit palaces —Manufactures, Education and Social Economy, Mines and Metallurgy, Varied Industries, Transportation, Liberal Arts, Agriculture, and Food Products—are practically completed and are ready for exhibits. The Palace of Machinery is ready. The steel work on the Palace of Fine Arts has been finished and the frame work for the rotunda is being placed. The Palace of Horticulture with its immense glass dome—152 feet in diameter and 180 feet high—has also been completed and the magnificent floral displays of the world are being planted therein. The steel Tower of Jewels, 433 feet high, is being constructed and will soon be finished. Three handsome fire stations are ready and automobile equipment has been installed in the central station. The million dollar

United States Fleet Steaming through the Golden Gate, Opposite Exposition Site

North Entrance, Palace of Machinery

North End Court of the Universe

Exposition Auditorium is nearing completion. The 4,000,000 square feet of road are being rapidly packed into smoothness and covered with resilient red rock which will be easy on the feet and will eliminate the glitter which tires the eyes. Hundreds of trees have been transplanted into their permanent places along the drives and in the gardens. More than ten thousand quick growing vines are spreading their tendrils along the walls of the buildings and the fence enclosing the Exposition, and millions of blossoming flowers have been planted in the gardens and courts. Hundreds of beautiful pieces of sculpture have been finished and placed in position and the immense canvases of the celebrated mural artists are ready to be placed. The concession area is being rapidly built up and is assuming the form of a great pleasure city. To date thirty-six foreign nations have accepted the invitation of the United States Government to participate and many of these have already commenced construction of their pavilions. Up to date forty-four States and Territories have either made provision for participation or are now actively engaged in doing so. From present indications practically every State will be represented and many of these are completing their magnificent buildings. The United States Government has appropriated $500,000 for the national exhibits at the Exposition and a bill is now before Congress with the recommendation of the President that $500,000 more be appropriated to erect a building in which to house them. Two hundred and twenty-six congresses and conventions have already chosen San Francisco as their meeting place in 1915. Applications have been received for space for many thousands of exhibits in the various exhibit palaces.

"We'll be ready, February 20, 1915," declared President Charles C. Moore of the Exposition two years ago, and that promise still holds good, backed by the good management of the Exposition directorate and the tireless energy of the builders of the mammoth celebration.

Ground Plan of the Panama-Pacific International Exposition

COLOR STUDIES
OF THE
EXPOSITION CITY

"Arch of the Rising Sun" at the Eastern Entrance to
the "Court of the Universe."

Part of the Great Central Court, the "Court of
the Universe."

Palace of Fine Arts and Glimpse of Lagoon Skirted
by Western Driveway.

Main Tower, or "Tower of Jewels," Rising to a
Height of 433 Feet, Studded with 125,000 "Jewels."

Perspective of the Main Exhibit Palace
site, in its combination of scenic beauty and
in the world. It is a natural amphitheatre co
hills, flanked by the wooded heights and fo
wonderful, blue, island-studded Bay of San
famous "Golden Gate." The Exposition Ci
ized dream of the best architectural genius
artists can do in color, all that modern sci
skilled gardeners and the California climate
live in the memory of beholders as long as me

, looking towards San Francisco Bay. The
practical advantages, is probably unequaled
vering 635 acres, backed by residence-covered
ortifications of the Presidio, fronting on the
Francisco, just inside the portals of the
y which covers these 635 acres is the real-
f America, supplemented by all that famous
nce can do in lighting effects and all that
can do in flowers and trees. Its beauty will
emory itself endures.

Tower and Great Cascade in the Form
of a Staircase, in the "Court of
Abundance," the East Central
Court of the Exposition.

Looking South in the "Court of Palms,"
Showing the Facade and Tower of the
Palace of Education, the Palace of
Horticulture in the Background.

The Palace of Horticulture, Built Almost Entirely of
Glass, and Covering Over Five Acres.

One of the most attractive and beautiful features of this Exposition will be the electrical illumination. By a system of flood lighting, a soft light will pervade the Courts at night, revealing the facades of the Palaces and the natural colors of the flowers. Concealed batteries of powerful projectors will cause tens of thousands of specially prepared glass "jewels," hung tremulous upon the towers, to flash like great diamonds, rubies, and emeralds. And searchlights of great power will weave in the night sky auroras of ever-changing color.

One of the Four Niches in the "Court of the Four
Seasons," the West Central Court.

San Francisco The Exposition City

> "And in that black, deserted zone
> They built a city, stone on stone;
> A city that, on history's page,
> Is crowned the marvel of its age."

SAN FRANCISCO—"The City Loved Around the World"—is at once representatively Western and the most cosmopolitan city in the world. The forty-niner crossing the plains by ox team walked its unpaved streets and the Spanish Padres conquering the burning deserts to the south established a mission and dreamed their dreams of conquest, campaigning from this point. Here, in an early

day, from around ocean, with snow-white sails all set, the peoples and products of other bay. These early settlers brought devotion that, as the years rolled character of a city that was in

San Francisco is typically metropolitan development its broadness of mind and pur phere and happy hospitality "Rugged West."

On a parallel of latitude Louis and ten miles north tered by mountains on the brilliant California sunshine San Francisco rightly boasts

the Horn and across an unknown swung through the Golden Gate lands to anchor in the spacious with them energy, courage and by, has become imbedded in the its very origin cosmopolitan.

Western, in that through all its citizens have maintained that pose, that care-free atmos- so often met with in the

about fifty miles south of St. of Richmond, Virginia, shel- north and east, with the tempered by cooling winds, of her invigorating climate.

The Union Ferry Depot   Entrance to San Francisco

Affiliated Colleges, San Francisco   Seal Rocks, San Francisco

Beginning in April and continuing until October the gentle trade winds blow over San Francisco. Coming directly from the ocean, these breezes carry with them the salt tang of the sea and are healthful and exhilarating.

The annual mean temperature of San Francisco is 56 degrees Fahrenheit. September is the warmest and January the coldest month. The mean temperature of September is 59.1 degrees and of January 49.2. In the last twenty years there have been only twenty-seven days during which the temperature exceeded 90 degrees, and in the same period it has not fallen below 32 degrees, the freezing point. The differences between day and night temperatures are small. The warmest hour, 2 P. M., has a mean temperature of 59.2, and the coolest hour, 6 A. M., has a mean temperature of 50.9 degrees. Such a climate admits of comfort to all who attend the Exposition.

To walk the long esplanade on the bay shore, the blood quickened by strengthening ocean airs, to rest in the balmy sunshine of the sheltered courts, to traverse the miles upon miles of enchanting aisles in the exhibit palaces in perfect physical comfort, will be one of the cherished experiences of a visit to the Panama-Pacific International Exposition. Those who come from tropic climes and from the heated sections of our own country should bring with them warm wraps.

Located in the center of the long coast strip, with an adequate rainfall and a large area of tributary territory, San Francisco maintains a confident and conservative attitude toward future growth and commercial importance. This feeling is reflected in the marvelous production of the Exposition at an initial cost to city and state of seventeen and a half millions of dollars, and that within half a decade after the recuperation from the great fire. It is most remarkable that a city that has spent in eight short years $375,000,000 in its renaissance—a sum equal to the

Along the Ocean Beach, San Francisco

Dutch Windmill

Japanese Tea Garden

Portals of the Past

Museum

Music Stand

Buffaloes

Conservatory

GOLDEN GATE
PARK

Naval Training Station, Goat Island

Along the Water Front, San Francisco

cost of the Panama Canal—should also build the largest, most beautiful, and, what promises to be the most successful, of world expositions.

San Francisco is a wealthy city. Her bank clearings for 1912 were $2,677,561,952, an amount almost equaling the combined clearings of the five next larger cities of the Coast, which were $2,690,516,590 for the same period. The assessed valuation for 1912 was $605,141,664. The assessed valuation per capita was $1,308.24, making this the wealthiest city on the Pacific Coast and the fifth wealthiest in the country.

In beauty of location and natural attractions San Francisco stands supreme among American cities. Situated upon the point of a peninsula, surrounded on three sides by ocean and bay, builded upon irregularly rising hills, with magnificent mountain and marine views on every hand, set in an infinity of earth, sea and sky, San Francisco charms the imagination and appeals to the soul.

Momentarily leaving the Exposition itself out of the question, the visitor in 1915 will find a world of interest and information in San Francisco that can not be duplicated elsewhere: the sylvan charm of Golden Gate Park with its Japanese tea gardens, buffalo and elk paddocks, museum, wonderful walks and drives, and beautiful gardens containing the products of two zones; a visit to the Cliff House and Seal Rocks; Sutro Heights; an automobile drive around the famous Ocean Boulevard or to one of the many beauty spots down the peninsula; a study of reconstructed San Francisco, with its Golden Gate, its splendid harbor, ocean frontage, wharves and shipping, parks, markets, military reservations, old Mission, public buildings, historic points and near-by resorts—the trip most interesting to the tourist is that through Chinatown, visiting the joss houses, the Chinese theaters, bazaars, curio stores, restaurants, markets, etc.— a visit to the Presidio, a sunny afternoon on Fisherman's Wharf or a lounge in one of the many beautiful parked squares that are found at

Kearny and Market Streets, San Francisco

Land's End, Outside Golden Gate

BUSINESS SECTION, SAN FRANCISCO

Market Street at Powell Street
SAN FRANCISCO

Market Street at Post Street
SAN FRANCISCO

Steamer Docks, San Francisco   Fisherman's Wharf, San Francisco

convenient intervals and serve as breathing places in the midst of the city's business and bustle; in the constant stir of cafe and hotel forming the city's night life—anywhere, everywhere, he will be impressed and thrilled with a feeling that here on the farthest shores of earth's greatest ocean the world is taking a holiday and he is part of it.

With San Francisco as the center a week or more can be well spent and at small cost in visiting the cities of Oakland, Alameda and Berkeley, a twenty-minute ride across the bay, and such near-by points of interest as Stanford University, the University of California at Berkeley, the Mare Island Navy Yard, Mill Valley, Mt. Tamalpais, the Muir Redwoods, Piedmont Springs, etc. The following are a few of the points of interest that can be reached from San Francisco, with the round trip fare in each case:

Oakland and Lake Merritt...................................$0.20
Idora Park, Oakland.. .20
Piedmont Springs, Oakland.................................. .20
University of California at Berkeley........................... .20
Sausalito, by steamer.. .25
Mill Valley.. .40
San Rafael.. .50
Steamer trip around San Francisco Bay......................... 1.00
Mare Island (the United States Navy Yard), by steamer.......... 1.00
The "Portola Discovery Trip" on the Ocean Shore Railroad, leaving San Francisco at 10 A. M. and returning at 5:15 P. M........... 1.00
Palo Alto for Stanford University, Sunday excursion, $1.05; two-day excursion 1.30
Trip to the base of Mount Diablo by ferry and electric railway. Round trip from San Francisco, week days, $2.10; Saturdays and Sundays 1.40
Mt. Tamalpais, over "the crookedest railroad in the world," and where a magnificent view can be had of the Pacific Ocean, of San Francisco Bay with its surrounding hills and mountains and of twenty-five cities... 1.90

Panorama of the New San Francisco

Curbstone Vendors, Chinatown   Dragon Procession, Chinatown

The Muir Woods via Mt. Tamalpais Railway (a grove of virgin Redwoods, some nearly 300 feet high, and within two hours' ride from San Francisco).................................... 1.90

The Mt. Tamalpais and Muir Woods trips can be combined in a day's outing for a round trip fare of............................. 2.90

The "Key-Trolley Trip," leaving San Francisco at 10 A. M. and 1 P. M., returning at 4:50 P. M., gives one sixty-eight miles of sight-seeing, visiting the University of California, the Greek Theater, the cities of Berkeley, Alameda, and Oakland, Piedmont Gardens and Springs and the Ostrich Farm. Fare for the round trip, including guide and admission to attractions............. 1.00

In the matter of public entertainment San Francisco can feel proud. She is second only to New York in the number and quality of her hotel accommodations. At present there are over 2000 hotels and apartment houses in San Francisco. This number is supplemented by the many up-to-date hostelries of the trans-bay cities of Oakland, Berkeley and Alameda. Very reasonable rates are in force and the hotel association has assured the Exposition officials that these rates will prevail during the Exposition period. Rooms occupied by one person may be obtained in San Francisco by the day from $1.00 up. Rooms with bath, $1.50 up. San Francisco is noted for the number and variety of her restaurants, where substantial meals can be obtained from 25 cents to $1.00. It is generally conceded that, quality for quality, the San Francisco restaurant prices are from 20 per cent to 40 per cent below those of New York City. The visitor to the Panama-Pacific International Exposition can be assured of the fact that he will not be overcharged by the hotels and restaurants. He will find that San Francisco is a city of hospitality and entertainment, and that its citizens are imbued with a desire to extend the right hand of fellowship and good will to the stranger within her gates in 1915.

Looking Towards the Bay of San Francisco

Chinatown
SAN FRANCISCO

California the Hostess

TAKE the sunniest parts of sunny Italy and Spain and the South of France with their wealth of vineyards and orchards; take the rugged mountain scenery of Switzerland and blend with it the verdure-clad hills of bonnie Scotland and the meadows and moors of rural England; place here and there the more beautiful bits of the French and Italian Rivieras with their wooded slopes and silvery beaches, joyous crowds, and gay life; bound this collection on one side by the earth's longest mountain range and on the other by the largest ocean, and cover with a canopy of turquoise blue sky and brilliant sunshine and you have a picture that yet falls short of—CALIFORNIA THE GOLDEN.

The name "California" is surrounded by the glamor and poetry of adventurous and romantic times—the advent of the Spanish don and conquistador, and their far from gentle acts, followed by the meek and loving mission of the good Father Serra, who, between the years 1769 and 1776, traveled over the hot sands, back and forth, for thousands of miles, and founded upwards of fifteen missions, establishing a practical Christianity which taught "Peace on earth, good will to all men."

The periods of the Spanish conquerors and the Christian conquest were followed, in 1848, by the wild stampede of the immigrants on the discovery of gold. This era has been made famous by the pens of Bret Harte, Mark Twain, and Joaquin Miller.

But the romantic of yesterday has given place to the practical of today. As the tourist rolls along over the beautifully smooth state highways in his high-powered car, he will only be reminded of past glories by an occasional glimpse of one of Father Serra's missions, which today, perchance, boasts a caretaker in place of a picturesque prior.

On the Beach at Santa Cruz

Donner Lake

Ostrich Tree, Seventeen-Mile Drive

From majestic Mount Shasta in the north to her sister, the picturesque Mount San Bernardino in the south—from the High Sierras to the shining sea—California abounds in scenery and opportunities wonderfully attractive to the tourist, the home-seeker, and the investor.

The climate of California is only one of her assets, but a very important one. To the salubrity of the climate can be attributed the virility and versatility of her native and adopted sons and daughters—writers, artists, sculptors, engineers, architects, scholars—who have brought fame to themselves and their beloved state in all parts of the earth, by reason of the out-of-doors-all-the-year-round climate that at all times enables one to sleep and eat with perfect enjoyment and work with rare diligence and a healthy ambition.

California has a land area of 155,980 square miles and a population of only about 2,500,000. With a temperate climate in the northern counties and almost tropical conditions in the south, she can boast of a diversity of products not equaled in any other part of the earth. She excels in dairying, cattle, and wheat growing, agriculture, horticulture, and viticulture. One county produces more raisins than the whole of Spain; one, more artichokes than the south of France; while yet another county produces more French prunes than the mother country, and the orange and lemon crop of California is greater than that of Europe. Everything produced in the Torrid or Temperate zones is grown—and grown to perfection—in California. The products of all the other states in the Union are duplicated here, together with many others, not grown elsewhere, but peculiar to the rich soil and kindly climate of California.

The mineral output of the state is another big asset. In fifty-five out of a total of fifty-eight counties minerals are found in paying quantities. Over one billion and a half in gold has been mined since 1848. The estimated mineral production for 1913 is one hundred million dollars.

Military Camp, Yosemite Meadows

Mount Tallac, Lake Tahoe

Moonlight on the Golden Gate

Yacht Club House, Sausalito   Tamalpais from Marin's Shores

Roughly speaking, California produces one-fourth of the world's output of oil, based on a total production of, approximately, 350,000,000 barrels.

The visitor to the Panama-Pacific International Exposition in 1915, however, will be interested in California's beautiful scenery, natural attractions, and places of renown, as well as in her commercial activity. Hence a short description of the principal places of interest.

The Missions. The missions of California are well worth a visit. They are scattered at intervals along the "Camino Real," or "Royal Highway," from San Diego to San Francisco. They are easy of access from the main thoroughfare and, by their peaceful setting and interesting inscriptions, invite the traveler to spend an hour or two "far from the madding crowd's ignoble strife." For the most part they are in a good state of preservation and vividly recall "the days before the Gringo came," when the Spaniards ruled the land and the Indians were their servants.

It is very hard to determine which is the most interesting of the missions. Dolores, in San Francisco, is the most important to the San Franciscan, as it gave the name to his city. The Mission Dolores was founded in 1776 and dedicated to San Francisco d'Assisi. It is very well preserved. A string of bells still hangs suspended by the original rawhide ropes. These are the bells that inspired Bret Harte to say:

> Bells of the past, whose long-forgotten music
>   Still fills the wide expanse,
> Tingeing the sober twilight of the present
>   With color of romance.
>
> I hear you call, and see the sun descending
>   On rock, and wave and sand,
> As down the Coast the mission voices blending
>   Girdle the heathen land.
>
> Borne on the swell of your long waves receding,
>   I touch the farther Past,—
> I see the dying glow of Spanish glory,
>   The sunset dream and last!

Oakland Sky Line from Beautiful Lake Merritt

Santa Ynez

San Juan Capistrano

Santa Barbara

San Luis Rey

San Fernando

Carmel

San Miguel

MISSIONS of CALIFORNIA

San Gabriel

Dolores

Miles of Artichokes   California, a Hunter's Paradise

> Before me rise the dome-shaped mission towers,
>   The white Presidio;
> The swart commander in his leathern jerkin,
>   The priest in stole of snow.
>
> Once more I see Portola's cross uplifting
>   Above the setting sun;
> And past the headland, northward, slowly drifting,
>   The freighted galleon.

Another very interesting mission is that of San Juan Bautista in the San Juan Valley of San Benito County. San Juan is eight miles west of Hollister, and about sixteen miles inland from the Bay of Monterey. It was founded in 1797 and has maintained its beauty of surroundings and sylvan seclusion through all the years. Of this mission the author of "Ramona" says:

"At San Juan Bautista there lingers more of the atmosphere of the olden time than is to be found in any other place in California.

"The mission church is well preserved; its grounds are enclosed and cared for; in its gardens are still blooming roses and vines, in the shelter of palms, and with the old stone sun dial to tell time.

"In the sacristy are oak chests, full of gorgeous vestments of brocades, with silver and gold laces. The church fronts south, on a little, green, locust-walled plaza—the sleepiest, sunniest, dreamiest place in the world."

Following is a list of the other important missions of California, together with their location:

San Francisco Solano Mission at Sonoma, San Rafael Archangel Mission at San Rafael, Santa Clara Mission at Santa Clara, Santa Cruz Mission at Santa Cruz, San Carlos de Borromeo Mission at Monterey, San Carlos de Rio Carmelo Mission at Monterey, Nuestra Senora de la Soledad Mission at Soledad, San Antonio de Padua Mission at King City, San Miguel Mission at San Miguel, San Luis Obispo de Tolosa Mission at San Luis Obispo, Santa Ynez Mission at Santa Ynez, La Purisima Concepcion Mission at Lompoc, Santa Barbara Mission at Santa Barbara, San Buenaventura Mission at Ventura, San Fernando Rey de Espana Mission at Fernando, San Gabriel Archangel Mission at Los Angeles, San

Polo Field—Golf and Polo Are Enjoyed the Year Around

Bohemian Grove

Grizzly Giant (MARIPOSA)

Fallen Monarch (MARIPOSA)

Mariposa Grove

BIG TREES OF
CALIFORNIA

Santa Cruz Grove

Vermont & Wawona (MARIPOSA)

An Orange Orchard   A California Home

Antonio de Pala Mission at Fallbrook, San Juan Capistrano Mission at
Capistrano, San Luis Rey de Francia Mission at Oceanside, San Diego de
Alcala Mission at San Diego, Santa Ysabel Mission at Foster.

The Great Central Valley. Between the two great mountain ranges
of California, the Sierra Nevada on the east and the Coast Range on the
west, lies the Great Central Valley, drained by the San Joaquin and the
Sacramento rivers. This valley extends from the Tehachapi Mountains
on the south to Mt. Shasta on the north, a distance of about 550 miles.
With nearly 20,000 square miles of comparatively level land, this is
both the granary of California and one of the great fruit and stock
producing regions of the world. The southern portion of the valley is .
known as the San Joaquin Valley, while the northern part is called the
Sacramento Valley. Visitors to the state wishing to study agricultural
California should by all means visit points in the "great valley." Here
is the home not only of grains, alfalfa, celery, and asparagus, but of the
fig, the almond, the grape, the orange, the apricot, the olive, and other
tropical and subtropical fruits.

Among the Redwoods. A week could be very pleasantly spent among
the redwoods at numerous hotels, mineral springs, or farm resorts north
of San Francisco. On this line special summer rates will be granted in
1915. A most attractive one-day jaunt over this line is the "Triangle
Trip," taking one through 150 miles of mountain and redwood forest
scenery, with a boat ride on San Francisco Bay, and by rail along the
Russian River. Round-trip rate for the "Triangle Trip": Sundays, $2.20;
Fridays and Saturdays, $2.50; other week days, $2.80. Hotel accommodations may be secured at these resorts at from $8.00 to $14.00 per week.

Calaveras Big Trees. A most interesting trip is that to the Calaveras
Big Trees, reached by rail from San Francisco via Stockton to Angels,
thence by stage twenty-two miles to the grove. This is the land of
Bret Harte and Mark Twain and of the placer mining of the days of '49.

A Typical Scene in the Oil Fields of California

Vista of Lake Tahoe · A California Pergola

The Sierra Road cuts through Table Mountain, recalling "Truthful James" and the "Society upon the Stanislaus." The route follows the famous Mother Lode, giving an opportunity to see something of deep quartz mining. Among other trips that can be taken at small cost are those to Mercer's Cave and to the Natural Bridge. Other side trips from this region are those to Lake Eleanor and the Hetch Hetchy Valley. The round trip fare from San Francisco to the Calaveras Big Trees is $14.60. Hotel accommodations from $12.00 per week up.

Shasta Resorts. All reaching San Francisco or returning home by the Shasta Route will find it well worth their while to stop over for a week or more at any one of the resorts near Mt. Shasta. Excursions to Mt. Shasta and to the numerous mineral springs, trips among the pines, mountain climbing, hunting and fishing are among the attractions of the Shasta region. Hotel accommodations from $12.00 per week up.

Lake Tahoe. During the open season, from May 15 to October 15, a week, or the entire vacation, for that matter, can be profitably spent at Lake Tahoe resorts (elevation 6,240 feet). Lake Tahoe is twenty-three miles long and thirteen miles wide. Those going to or returning from San Francisco may stop over at Truckee and visit the Tahoe resorts at but little extra expense for side-trip transportation. Stop-overs at Truckee will be allowed on all through railway and Pullman tickets. A round trip ticket from Truckee to the lake, around the lake by "Steamer Tahoe," and return to Truckee will cost $6.00. On this ticket stop-overs will also be allowed. Among Tahoe amusements are trout fishing in the lake and numerous streams found round about, bathing, boating, driving, and mountain climbing. Accommodations may be secured at the Tahoe resorts at from $2.00 per day up.

Yosemite National Park. Those reaching San Francisco via the San Joaquin Valley or with return tickets via the San Joaquin Valley can

Rubidoux Cross and Mt. San Bernardino · Mossbrae Falls, Shasta Country

On the Campus   University of California   Le Conte Oak

arrange to reach the Yosemite National Park from Merced, all tickets permitting stop-over privileges at Merced. From Merced the round-trip rate to the Park is $18.50. For those not routed via the San Joaquin Valley, the round trip rate to the Yosemite National Park from San Francisco will be $22.35 for those traveling on day trains, with $2.00 each way added for Pullman for those taking the night train. Camp accommodations in the park can be secured at from $2.50 to $3.00 per day; hotel rates from $3.50 to $5.00 per day. Tents for private camping may be rented at reasonable rates. Trained saddle horses may be hired in the park at from $2.50 to $4.00 per day. Many tourists in the park take the trails on foot, thus eliminating the expense for saddle horses as well as securing the enjoyment of mountain climbing. Twenty-six miles from the park is the Mariposa Grove of Big Trees. This grove can be reached by stage, the round trip costing $15.00. From El Portal, the Merced Grove of Big Trees may be reached by stage at an expense of not to exceed $7.50.

The Canyons. A pleasant week may be spent in the Sequoia National Park east of Visalia, or in the neighboring canyons of the Kings and Kern Rivers, which, with their higher surrounding mountains, offer attractions only equaled by the Yosemite. Those going to San Francisco or returning via the San Joaquin Valley may stop over at Visalia or Exeter. The round trip from there to Camp Sierra in the Giant Forest, where are the greatest number of Big Trees in the world, is about $13.00, including electric railway and stage ride. In the Giant Forest are more than 3,000 Big Trees over 300 feet high, with many thousands more of lesser size. The round trip from Visalia or Exeter, including a week's accommodations at Camp Sierra, transportation, etc., would be about $25.00. The rate at Camp Sierra for tent and board is $2.00 per day or $50.00 per month.

Bay at Santa Catalina Island   In the Gold Lake Country

Cathedral Rock
YOSEMITE

Main Gateway, Stanford University   Lick Observatory, Mt. Hamilton

Coast Resorts. A week and as much longer as one wishes to remain could be delightfully spent at Santa Cruz, Monterey, Del Monte, Pacific Grove, Paso Robles Hot Springs, or El Pizmo Beach, resorts between San Francisco and Los Angeles. Among the attractions at either Santa Cruz or Pacific Grove are surf-bathing, boating, and fishing. Between Pacific Grove and Monterey, one may journey by street car, take the famous Seventeen-Mile Drive, visiting Carmel Mission, etc. On all railroad tickets stop-overs will be allowed at Palo Alto to visit Stanford University (one mile from Palo Alto), and at San Jose to inspect the orchards of the Santa Clara Valley or to visit the Lick Observatory on Mount Hamilton, 4209 feet elevation (round trip by stage, $5.00). Accommodations may be secured at the Coast resorts at from $12.00 per week up.

Santa Cruz Big Trees. The Santa Cruz Grove of Big Trees is seventy miles south of San Francisco and six miles north of Santa Cruz. The trees in this grove are known as Sequoia sempervirens or Redwood. The "Giant," the largest tree in the grove, is 64 feet in circumference and 306 feet high. The grove may be reached from Santa Cruz by automobile or tally-ho or via the railroad. Nineteen miles from Felton and twelve miles from Boulder Creek, is California Redwood Park, a state park of 3,800 acres of natural forest.

The Santa Clara Valley. The orchards of the Santa Clara Valley may be toured from San Jose by automobile or tally-ho. The entire western section of the valley may be seen from the trolley cars. A forty-mile ride over this line may be made between San Jose and Palo Alto for 90 cents, while a journey over the entire line, returning to starting point, may be taken on the Blossom Trolley Trips by cars which leave San Jose, Los Gatos, and Palo Alto every day between 9:30 and 10:30 A. M. for $1.00. From Palo Alto cars run every ten minutes to Stanford University. Alum Rock Canyon, the unique city park of San Jose, may be reached by cars leaving the center of the city; fare 10 cents each way.

Palm Canyon—"Arabia" in California   A Suburban Roadway

Nevada Falls

View from Moran Point

Yosemite Falls YOSEMITE NATIONAL PARK El Capitan

State Capitol, Sacramento   An Oakland Hillside Home

Santa Barbara. A week in Santa Barbara would give a most delightful rest. A visit to the old mission, the beach, the many drives and trails, will all prove of the greatest interest. Hotel accommodations from $12.00 per week up.

Los Angeles and Southern California. From Los Angeles many delightful and inexpensive trips can be made to San Diego and other points of interest throughout Southern California. The cost for room and meals in Los Angeles, San Diego, Catalina, Santa Monica, Long Beach, Redondo, and other nearby resorts will be about the same as in San Francisco. The following are a few of the points of interest that may be reached from Los Angeles, with the round trip rate in each case:

Pasadena and the Ostrich Farm.................................$ .25
Santa Monica, Redondo Beach, Long Beach, Venice or San Pedro.. .50
"Seeing Los Angeles" by auto or observation car.................. .50
Old Mission Trolley Trip, including Pasadena, Baldwin's Ranch,
   Monrovia, San Gabriel Mission, and Alhambra................ 1.00
Balloon Route Trolley Trip, traveling thirty-six miles along the
   ocean shore, visiting ten beaches and eight cities............. 1.00
Triangle Trolley Trip, visiting Santa Ana, Huntington Beach, Naples,
   Long Beach, Point Fermin, and San Pedro................... 1.00
Mount Lowe Trolley Trip, through Pasadena and Rubio Canyon,
   Echo Mountain and Alpine Tavern.......................... 2.50
Santa Catalina, the island resort, 60-day ticket, $2.75; ticket Saturday
   and Sunday with return limit on Monday..................... 2.50
The "Kite-Shaped Track" trip, the "Inside Track" trip, or the
   "Orange Belt" trip.. 3.00

Feather River Canyon. Those reaching San Francisco and the Exposition over this route will have an opportunity of seeing the rock-walled canyons of the Feather River. In this section are many resorts, with near-by streams well stocked with trout. The scenery is grand and beautiful, and deer, bear and smaller game are plentiful. Hotel accommodations from $10.00 per week up.

Greek Theatre, Berkeley, Showing Part of Audience of 8000 Persons

SCENIC CALIFORNIA

Eel River at Shirley

Devil's Elbow and Eel River

The Junipero Oak planted at Monterey 1770.

Palm Cañon near Palm Spring

Automobile Speedway, Corona

Midway Point, Near Monterey

Automobiling. From San Francisco as a center the motorist in 1915 may reach all parts of the Coast over smooth, well constructed state and county highways. The people of California have voted eighteen million dollars to build two highways from north to south, one through the great interior valley, the other along the coast. Much of the work has already been done and the remainder, it is hoped, will be completed by 1915. By order of the Secretary of the Interior, the ban against entering Yosemite Valley with automobiles has been removed. The automobile interests of the country are agitating the construction of a transcontinental highway from the Atlantic to the Pacific. At the time this book goes to press the success of this project is assured and the motorist in 1915 will find a concrete road across the country, the western end of which touches San Francisco. The Exposition, with the assistance of the automobile clubs, is marking all the good roads leading into this highway with blue and white embossed steel signs, so that the tourist from any part of the country, by following these signs, will be directed into the transcontinental road by the quickest and easiest route.

Hunting and Fishing. California has long been known as a Paradise for the disciples of Nimrod and Izaak Walton. Its forested mountains are the haunts of deer, bear, California lion, grouse, quail, and other game, and its streams are full of fish. Lake and brook trout are abundant in the mountain streams, and bass, salmon, and shad are the favorites in the valley. Along the coast there is an infinite variety of sport, from casting with ordinary rod and line to heroic struggles with gigantic tuna. Millions of wild geese swarm the fields in the interior in fall and spring, and the marshes and sheltered streams of the Great Valley are hunting grounds for wild ducks. For those who prefer to hunt with the camera, a trip to the Yosemite National Park, where the use of firearms is forbidden, and where deer may often be surprised, affords rare sport. Good hunting or fishing grounds in the mountains may be reached by rail, from San Francisco, with short trips by stage or on foot into the wilds.

Double Natural Bridge, Santa Cruz

MAP SHOWING SIDE TRIPS
TAKING SAN FRANCISCO AS A CENTER

NIGHT VIEW
SHOWING ILLUMINATION OF
SOUTH GARDENS AND MAIN ENTRANCE
PANAMA PACIFIC INTERNATIONAL EXPOSITION
SAN FRANCISCO 1915

"One of the most attractive and beautiful features of this Exposition will be the electrical illumination. By an entirely new system of flood lighting a soft, restful, yet perfect light will pervade the courts at night, revealing in wonderful clearness the facades and walls of the palaces and the natural colors of the shrubbery and flowers. By peculiar and novel lighting devices the statuary and mural paintings will be made to appear with even heightened effect. Concealed batteries will project powerful yet softened rays of light that will cause tens of thousands of specially prepared glass "jewels," hung tremulously upon the towers, to flash and scintillate like great diamonds, emeralds and rubies. At a point on the bay shore will be erected apparatus that will weave in the night sky auroras of ever-changing color. Altogether the spectacle will be interesting and wonderful and never to be forgotten."